Visit Cram101.com for full Practice Exams

Textbook Outlines, Highlights, and Practice Quizzes

Waves, Tides, and Shallow Water Processes

by Open University Team, 2nd Edition

All "Just the Facts101" Material Written or Prepared by Cram101 Publishing

Title Page

cram101.com

"Just the Facts101" is a Cram101 publication and tool designed to give you all the facts from your textbooks. Visit Cram101.com for the full practice test for each of your chapters for virtually any of your textbooks.

Cram101 has built custom study tools specific to your textbook. We provide all of the factual testable information and unlike traditional study guides, we will never send you back to your textbook for more information.

YOU WILL NEVER HAVE TO HIGHLIGHT A BOOK AGAIN!

Cram101 StudyGuides

All of the information in this StudyGuide is written specifically for your textbook. We include the key terms, places, people, and concepts... the information you can expect on your next exam!

Want to take a practice test?

Throughout each chapter of this StudyGuide you will find links to cram101.com where you can select specific chapters to take a complete test on, or you can subscribe and get practice tests for up to 12 of your textbooks, along with other exclusive cram101.com tools like problem solving labs and reference libraries.

Cram101.com

Only cram101.com gives you the outlines, highlights, and PRACTICE TESTS specific to your textbook. Cram101.com is an online application where you'll discover study tools designed to make the most of your limited study time.

By purchasing this book, you get 50% off the normal subscription free!. Just enter the promotional code **'DK73DW14042'** on the Cram101.com registration screen.

www.Cram101.com

Learning System

Waves, Tides, and Shallow Water Processes
Open University Team, 2nd

CONTENTS

CHAPTER OUTLINE: KEY TERMS, PEOPLE, PLACES, CONCEPTS

	Estuary
	Tidal power
	Mudflat
	Wave height
	Blake Plateau
	S-wave
	Canary Current
	Current
	Capillary wave
	Rossby wave
	Internal wave
	Pycnocline
	Vorticity
	Wind stress
	Significant wave height
	Wind wave
	Dispersion
	Persian Gulf
	Agulhas Current

Wave power

Pleistocene

Seiche

Standing wave

Tsunami

Resonance

Quaternary

Remote sensing

China Seas

Tidal range

Sediment

Lunar month

Kelvin wave

North Sea

Baltic Sea

Black Sea

Stratification

Fish migration

Continental shelf

CHAPTER OUTLINE: KEY TERMS, PEOPLE, PLACES, CONCEPTS

	Bay of Bengal
	Storm surge
	Tidal bore

CHAPTER HIGHLIGHTS & NOTES: KEY TERMS, PEOPLE, PLACES, CONCEPTS

Estuary	An estuary is a partly enclosed coastal body of water with one or more rivers or streams flowing into it, and with a free connection to the open sea. Estuaries form a transition zone between river environments and ocean environments and are subject to both marine influences, such as tides, waves, and the influx of saline water; and riverine influences, such as flows of fresh water and sediment. The inflow of both seawater and freshwater provide high levels of nutrients in both the water column and sediment, making estuaries among the most productive natural habitats in the world.
Tidal power	Tidal power is a form of hydropower that converts the energy of tides into useful forms of power - mainly electricity. The first study of large scale tidal power plants was by the US Federal Power Commission in 1924 which would have been located if built in the northern border area of the US state of Maine and the south eastern border area of the Canadian province of New Brunswick, with various dams, powerhouses and ship locks enclosing the Bay of Fundy and Passamaquoddy Bay . Nothing came of the study and it is unknown whether Canada had been approached about the study by the US Federal Power Commission.
Mudflat	Mudflats are coastal wetlands that form when mud is deposited by tides or rivers. They are found in sheltered areas such as bays, bayous, lagoons, and estuaries. Mudflats may be viewed geologically as exposed layers of bay mud, resulting from deposition of estuarine silts, clays and marine animal detritus. Most of the sediment within a mudflat is within the intertidal zone, and thus the flat is submerged and exposed approximately twice daily.
Wave height	In fluid dynamics, the wave height of a surface wave is the difference between the elevations of a crest and a neighbouring trough.

Wave height is a term used by mariners, as well as in coastal, ocean and naval engineering.

At sea, the term significant wave height is used as a means to introduce a well-defined and standardized statistic to denote the characteristic height of the random waves in a sea state.

Blake Plateau

The Blake Plateau lies between the continental shelf and the deep ocean basin 145 kilometers (90 miles) and 170 kilometers (106 miles) with a depth of about 500 meters (1,640 feet) inshore sloping to about 1,000 meters (3,281 feet) about 375 kilometers (233 miles) off shore where the Blake Escarpment drops steeply to the deep basin. Blake Plateau lies off the southeastern United States coastal states of North Carolina, South Carolina, Georgia and Florida. The Blake Plateau, associated ridge and basin are named for the U.S. Coast and Geodetic Survey steamer George S. Blake, in service 1874-1905, that first used steel cable for oceanographic operations and pioneered deep ocean and Gulf Stream exploration.

S-wave

A type of seismic wave, the S-wave, secondary wave, or shear wave (sometimes called an elastic S-wave) is one of the two main types of elastic body waves, so named because they move through the body of an object, unlike surface waves.

The S-wave moves as a shear or transverse wave, so motion is perpendicular to the direction of wave propagation: S-waves are like waves in a rope, as opposed to waves moving through a slinky, the P-wave. The wave moves through elastic media, and the main restoring force comes from shear effects.

Canary Current

The Canary Current is a wind-driven surface current that is part of the North Atlantic Gyre. This eastern boundary current branches south from the North Atlantic Current and flows southwest about as far as Senegal where it turns west and later joins the Atlantic North Equatorial Current. he Canary Islands.

Current

A current, in a river or stream, is the flow of water influenced by gravity as the water moves downhill to reduce its potential energy. The current varies spatially as well as temporally within the stream, dependent upon the flow volume of water, stream gradient, and channel geometrics. In tidal zones, the current in rivers and streams may reverse on the flood tide before resuming on the ebb tide.

Capillary wave

A capillary wave is a wave traveling along the phase boundary of a fluid, whose dynamics are dominated by the effects of surface tension.

Capillary waves are common in nature and the home, and are often referred to as ripples. The wavelength of capillary waves in water is typically less than a few centimeters.

Rossby wave	Atmospheric Rossby waves are giant meanders in high-altitude winds that are a major influence on weather. They are not to be confused with oceanic Rossby waves, which move along the thermocline: that is, the boundary between the warm upper layer of the ocean and the cold deeper part of the ocean. Rossby waves are a subset of inertial waves.
Internal wave	Internal waves are gravity waves that oscillate within, rather than on the surface of, a fluid medium. They are one of many types of wave motion in stratified fluids (another example being Lee waves). A simple example is a wave propagating on the interface between two fluids of different densities, such as oil and water.
Pycnocline	A pycnocline is the cline or layer where the density gradient ($\partial \rho/\partial z$) is greatest within a body of water. An ocean current is generated by the forces such as breaking waves, terms of temperature and salinity differences, wind, Coriolis effect, and tides caused by the gravitational pull of the Moon and the Sun. In addition, the physical properties in pycnocline driven by density gradients also affect the flows and vertical profiles in the ocean.
Vorticity	Vorticity is a concept used in fluid dynamics. In the simplest sense, vorticity is the tendency for elements of the fluid to 'spin.' More formally, vorticity can be related to the amount of 'circulation' or 'rotation' (or more strictly, the local angular rate of rotation) in a fluid. The average vorticity ω_{av} in a small region of fluid flow is equal to the circulation Γ around the boundary of the small region, divided by the area A of the small region.
Wind stress	In physical oceanography and fluid dynamics, the wind stress is the shear stress exerted by the wind on the surface of large bodies of water - such as oceans, seas, estuaries and lakes. It is the force component parallel to the surface, per unit area, as applied by the wind on the water surface. The wind stress is affected by the wind speed, the shape of the wind waves and the atmospheric stratification.
Significant wave height	In physical oceanography, the significant wave height is defined traditionally as the mean wave height (trough to crest) of the highest third of the waves ($H_{1/3}$), but now usually defined as four times the standard deviation of the surface elevation (or equivalently as four times the square root of the first moment of the wave spectrum). The symbol H_{m0} is usually used for that latter definition. The significant wave height may thus refer to H_{m0} or $H_{1/3}$, the difference in magnitude between the two definition is only a few percent.

Wind wave	In fluid dynamics, wind waves or, more precisely, wind-generated waves are surface waves that occur on the free surface of oceans, seas, lakes, rivers, and canals or even on small puddles and ponds. They usually result from the wind blowing over a vast enough stretch of fluid surface. Waves in the oceans can travel thousands of miles before reaching land.
Dispersion	In fluid dynamics, dispersion of water waves generally refers to frequency dispersion, which means that waves of different wavelengths travel at different phase speeds. Water waves, in this context, are waves propagating on the water surface, and forced by gravity and surface tension. As a result, water with a free surface is generally considered to be a dispersive medium.
Persian Gulf	The Persian Gulf, in Western Asia, is an extension of the Indian Ocean located between Iran (Persia) and the Arabian Peninsula. The Persian Gulf was the focus of the 1980-1988 Iran-Iraq War, in which each side attacked the other's oil tankers. In 1991, the Persian Gulf again was the background for what was called the 'Persian Gulf War' or the 'Gulf War' when Iraq invaded Kuwait and was subsequently pushed back, despite the fact that this conflict was primarily a land conflict.
Agulhas Current	The Agulhas Current is the Western Boundary Current of the southwest Indian Ocean. It flows down the east coast of Africa from 27°S to 40°S. It is narrow, swift and strong. It is even suggested that the Agulhas is the largest western boundary current in the world ocean, as comparable western boundary currents transport less, ranging from the Brazil Current, 16.2 Sverdrups), to the Kuroshio, 42 Sverdrups.
Wave power	Wave power is the transport of energy by ocean surface waves, and the capture of that energy to do useful work -- for example, electricity generation, water desalination, or the pumping of water (into reservoirs). Machinery able to exploit wave power is generally known as a wave energy converter (WEC).
Pleistocene	The Pleistocene (symbol P_S) is the epoch from 2,588,000 to 11,700 years BP that spans the world's recent period of repeated glaciations. The name pleistocene is derived from the Greek πλε?στος (pleistos 'most') and καιν?ς (kainos 'new'). The Pleistocene Epoch follows the Pliocene Epoch and is followed by the Holocene Epoch.
Seiche	A seiche is a standing wave in an enclosed or partially enclosed body of water. Seiches and seiche-related phenomena have been observed on lakes, reservoirs, swimming pools, bays, harbors and seas. The key requirement for formation of a seiche is that the body of water be at least partially bounded, allowing the formation of the standing wave.

Standing wave	In physics, a standing wave - also known as a stationary wave - is a wave that remains in a constant position.
	This phenomenon can occur because the medium is moving in the opposite direction to the wave, or it can arise in a stationary medium as a result of interference between two waves traveling in opposite directions. In the second case, for waves of equal amplitude traveling in opposing directions, there is on average no net propagation of energy.
Tsunami	A tsunami is a series of water waves caused by the displacement of a large volume of a body of water, typically an ocean or a large lake. Earthquakes, volcanic eruptions and other underwater explosions (including detonations of underwater nuclear devices), landslides, glacier calvings, meteorite impacts and other disturbances above or below water all have the potential to generate a tsunami.
	Tsunami waves do not resemble normal sea waves, because their wavelength is far longer.
Resonance	In physics, resonance is the tendency of a system to oscillate at a greater amplitude at some frequencies than at others. These are known as the system's resonant frequencies (or resonance frequencies). At these frequencies, even small periodic driving forces can produce large amplitude oscillations, because the system stores vibrational energy.
Quaternary	The Quaternary Period is the most recent of the three periods of the Cenozoic Era in the geologic time scale of the ICS. It follows the Neogene Period, spanning 2.588 ± 0.005 million years ago to the present. The Quaternary includes two geologic epochs: the Pleistocene and the Holocene. Research history
	The term Quaternary was proposed by Giovanni Arduino in 1759 for alluvial deposits in the Po river valley in northern Italy.
Remote sensing	Remote sensing is the acquisition of information about an object or phenomenon, without making physical contact with the object. In modern usage, the term generally refe to the use of aerial sensor technologies to detect and classify objects on Earth (both on the surface, and in the atmosphere and oceans) by means of propagated signals (e.g. electromagnetic radiation emitted from aircraft or satellites).
	There are two main types of remote sensing: passive remote sensing and active remote sensing.

Chapter 1. PART I: Chapter 1 - Chapter 2

Tidal range	The tidal range is the vertical difference between the high tide and the succeeding low tide. Tides are the rise and fall of sea levels caused by the combined effects of the gravitational forces exerted by the Moon and the Sun and the rotation of the Earth. The tidal range is not constant, but changes depending on where the sun and the moon are.
Sediment	Sediment, a naturally occurring material that is broken down by processes of weathering and erosion, and is subsequently transported by the action of fluids such as wind, water, or ice, and/or by the force of gravity acting on the particle itself. Sediments are most often transported by water (fluvial processes), wind (aeolian processes) and glaciers. Beach sands and river channel deposits are examples of fluvial transport and deposition, though sediment also often settles out of slow-moving or standing water in lakes and oceans.
Lunar month	In lunar calendars, a lunar month is the time between two identical syzygies (new moons or full moons). There are many variations. In Middle-Eastern and European traditions, the month starts when the young crescent moon becomes first visible at evening after conjunction with the Sun 1 or 2 days before that evening (e.g. in the Islamic calendar).
Kelvin wave	A Kelvin wave is a wave in the ocean or atmosphere that balances the Earth's Coriolis force against a topographic boundary such as a coastline, or a waveguide such as the equator. A feature of a Kelvin wave is that it is non-dispersive, i.e., the phase speed of the wave crests is equal to the group speed of the wave energy for all frequencies. This means that it retains its shape in the alongshore direction over time.
North Sea	The North Sea is a marginal sea of the Atlantic Ocean located between Great Britain, Scandinavia, Germany, the Netherlands, and Belgium. An epeiric (or 'shelf') sea on the European continental shelf, it connects to the ocean through the English Channel in the south and the Norwegian Sea in the north. It is more than 970 kilometres (600 mi) long and 580 kilometres (360 mi) wide, with an area of around 750,000 square kilometres (290,000 sq mi).
Baltic Sea	The Baltic Sea is a brackish mediterranean sea located in Northern Europe, from 53°N to 66°N latitude and from 20°E to 26°E longitude. It is bounded by the Scandinavian Peninsula, the mainland of Europe, and the Danish islands. It drains into the Kattegat by way of the Øresund, the Great Belt and the Little Belt.
Black Sea	The Black Sea is bounded by Europe, Anatolia and the Caucasus and is ultimately connected to the Atlantic Ocean via the Mediterranean and the Aegean seas and various straits. The Bosphorus strait connects it to the Sea of Marmara, and the strait of the Dardanelles connects that sea to the Aegean Sea region of the Mediterranean.

Stratification	Water stratification occurs when water of high and low salinity (halocline), oxygenation (chemocline), density (pycnocline), temperature (thermocline), forms layers that act as barriers to water mixing.
Fish migration	Many types of fish migrate on a regular basis, on time scales ranging from daily to annually or longer, and over distances ranging from a few metres to thousands of kilometres. Fish usually migrate because of diet or reproductive needs, although in some cases the reason for migration remains unknown.

Classifications can be either fundamental (like biological classification that rests on a phylogenetic basis), or are merely heuristic typologies (as here, about migrations) to assist communication about complex issues. Classifications are judged according to their fundamental accuracy, whether they are convenient or not. Typologies in contrast are essentially arbitrary and their effectiveness is to be judged solely by the problems they solve or create. Secor and Kerr (2009) for example show several typologies that encapsulate various aspects of fish life history.

Migration is a word used in multiple senses. It is important to distinguish 'true' migration, i.e. a life-history-structured or at least patterned activity such as seen in anadromous species like salmons, from mere movement or wandering as may happen, say, with euryhaline species that easily move between fresh and salt water but not necessarily or with regularity. 'Vertical migration', for example, the phenomenon of plankton and fishes regularly changing their depth throughout the 24h day, is a special usage unlike migrations of (e.g) salmons that range over distances in migrations that may cover river, lake, and sea, or the great migrations of game through Africa's Serengeti.

Anadromous and catadromous are words that have been commonly used for centuries. They are slightly more narrowly used in the following classification of [truly] migrating fish by Myers 1949:'Diadromous. Truly migratory fishes which migrate between the sea and fresh water. There has been no English term by which one can refer collectively and briefly to anadromous, catadromous and other fishes which truly migrate between fresh and salt water, and this new term is now proposed ... . Like the two well known ones, this adjective is formed from classical Greek ([dia], through; and [dromous], running). ...Anadromous. Diadromous fishes which spend most of their lives in the sea and migrate to fresh water to breed (From [ana], up .)... ... [This narrowed the previous usage which could apply to fish never crossing the freswater/sea boundary but simply moving upstream to spawn, for example in some fishes of the Rift Lakes]Catadromous. Diadromous fishes which spend most of their lives in fresh water and migrate to the sea to breed. (From [cata], down .)... ... [This narrowed the previous usage which could apply to fish never reaching the sea but moving downstream to spawn]Amphidromous. Diadromous fishes whose migration from fresh water to the seas, or vice versa, is not for the purpose of breeding, but occurs regularly at some other stage of the life cycle.

(From [amphi], around, on both sides .).. ... [The elements 'vice versa' and 'purpose' are troublesome however]Potamodromous. Truly migratory fishes whose migrations occur wholly within freshwater. (From [potamos], river ...) ... [Rarely used. This term and oceanodromous received the fishes excluded by the narrowing of anadromous and catadromous]Oceanodromous. Truly migratory fishes which live and migrate wholly in the sea. (From [oceanos], the ocean ...) ... [Rarely used. This term and potamodromous received the fishes excluded by the narrowing of anadromous and catadromous] '

The '-ous' endings are for the adjectival form of the terms; nouns are obtained by replacing that ending with '-y', e.g. anadromy.

The terms anadromous and catadromous were of long standing (and similar but not identical usage); the other terms were coined by Myers. Myers was hesitant about introducing new terms however, saying:'The writer's strong aversion to the infliction of new scientific terms upon the scientific public has, however, caused him to proceed with great circumspection. ... Finally, the writer has ventured to propose new terms only because he feels that certain new terms will be of distinct advantage in some specialized types of ichthyological work, and that the general biologist and fishery worker will seldom or never have to bother his head about them. Catadromous and anadromous will almost certainly always remain the most important and probably the only widely used terms of their class.'(emphasis added)

Myers' term diadromous has proved useful as an inclusive term. But his prediction was otherwise accurate: anadromous and catadromous, are indeed the main widely understood terms. Secor and Kerr (2009) found only one paper that used oceanodromous, 18 using potamodromous, and 122 using amphidromous compared to 985 times for anadromous, 143 times for diadromous, 71 times for catadromous. Obviously the amount of work done on a group is one source of this variation, but another source is the perceived utility and standing of the term itself.

Amphidromy is rarely understood by the wider fish biology audience, or alternatively 'There seems to be some reluctance to use the term amphidromy'. Myers' definition's ambiguity ('or vice versa') and teleology ('for the purpose of') is one part of that problem. Its meaning/usage has been adjusted (but two such adjustments were found zero times by Secor and Kerr (2009)), and recently it has been stated as differing from anadromy simply because return to fresh water occurs at a juvenile or immature stage; other differences have been claimed (such as where most growth occurs) but they are necessary consequences of the stage at return and therefore are not independent. Some authors have therefore eschewed amphidromy in favour of the more widely understood terms: either anadromy, with or without a remark that return to rivers occurs at an earlier stage (e.g. 'juvenile-return anadromy'), or diadromy which discards information that would be conveyed by anadromy with or without remark.

It has to be borne in mind that any typological system has heuristic value as a convenience for description, and need not reflect phylogeny: each category (term) may include representatives of many distantly related taxa, each of which may well have close relatives that are in another category (term) or that are not migratory and thus fall completely outside the typology. These terms are therefore of limited safety when used alone to screen data to be considered for an analysis, or to decide which species should be read about to explore a phenomenon or possible comparison.

The limitations of typologies were clearly stated by Myers himself. Each may be useful in one context, for one purpose, but not another. I.e. they may categorise along different axes (see Secor and Kerr (2009)). For example, Myers in the same year devised another interesting set of terms, also directed at fishes, but on the basis of their salt-tolerance.

As a footnote it is interesting to note that George S. Myers had previously been one of the degree supervisors for Porfirio Manacop, whose ground-breaking Master's work (in 1941, later published 1953) on a Sicyopterus species overturned the prevailing notion that it was catadromous. It is likely that Myers was impressed by this, and used Sicydium as the 'type' genus for amphidromous.

And, as a contrary footnote showing how science fails: regrettably, Manacop's work seems to have been as locally unpopular as it was innovative, because over 20 years later the group he worked on was still being referred to by a colleague in the Philippines who would certainly have known of his work, incorrectly and without evidence, as catadromous.

And although these systems were originated for fishes, they are in principle applicable to any organism. Forage fish

Forage fish often make great migrations between their spawning, feeding and nursery grounds. Schools of a particular stock usually travel in a triangle between these grounds. For example, one stock of herrings have their spawning ground in southern Norway, their feeding ground in Iceland, and their nursery ground in northern Norway. Wide triangular journeys such as these may be important because forage fish, when feeding, cannot distinguish their own offspring.

Capelin are a forage fish of the smelt family found in the Atlantic and Arctic oceans. In summer, they graze on dense swarms of plankton at the edge of the ice shelf. Larger capelin also eat krill and other crustaceans. The capelin move inshore in large schools to spawn and migrate in spring and summer to feed in plankton rich areas between Iceland, Greenland, and Jan Mayen. The migration is affected by ocean currents. Around Iceland maturing capelin make large northward feeding migrations in spring and summer. The return migration takes place in September to November.

The spawning migration starts north of Iceland in December or January.

The diagram on the right shows the main spawning grounds and larval drift routes. Capelin on the way to feeding grounds is coloured green, capelin on the way back is blue, and the breeding grounds are red The Convention does not provide an operational definition of the term, but in an annex (UNCLOS Annex 1) lists the species considered highly migratory by parties to the Convention. The list includes: tuna and tuna-like species (albacore, bluefin, bigeye tuna, skipjack, yellowfin, blackfin, little tunny, southern bluefin and bullet), pomfret, marlin, sailfish, swordfish, saury and ocean going sharks, dolphins and other cetaceans.

These high trophic oceanodromous species undertake migrations of significant but variable distances across oceans for feeding, often on forage fish, or reproduction, and also have wide geographic distributions. Thus, these species are found both inside the 200 mile exclusive economic zones and in the high seas outside these zones. They are pelagic species, which means they mostly live in the open ocean and do not live near the sea floor, although they may spend part of their life cycle in nearshore waters.

Highly migratory species can be compared with straddling stock and transboundary stock. Straddling stock range both within an EEZ as well as in the high seas. Transboundary stock range in the EEZs of at least two countries. A stock can be both transboundary and straddling. Other examples

Some of the best-known anadromous fish are the six species of Pacific salmon, which are Chinook (King), Coho (Silver), Sockeye (Red), Chum (Dog), Pink (Humpback), and Cherry. The salmon hatch in small freshwater streams. From there they migrate to the sea to mature, living there for two to six years. When mature, the salmon return to the same streams where they were hatched to spawn. Salmon are capable of going hundreds of kilometers upriver, and humans must install fish ladders in dams to enable the salmon to get past. Other examples of anadromous fishes are sea trout, three-spined stickleback, and shad.

The most remarkable catadromous fishes are freshwater eels of genus Anguilla, whose larvae drift from swawning grounds in the Sargasso sea, sometimes for months or years, before entering freshwater river and streams as glass eels or elvers .

An example of a euryhaline species is the Bull shark, which lives in Lake Nicaragua of Central America and the Zambezi River of Africa. Both these habitats are fresh water, yet Bull sharks will also migrate to and from the ocean. Specifically, Lake Nicaragua Bull sharks migrate to the Atlantic Ocean and Zambezi Bull sharks migrate to the Indian Ocean.

Diel vertical migration is a common behavior; many marine species move to the surface at night to feed, then return to the depths during daytime.

A number of large marine fishes, such as the tuna, migrate north and south annually, following temperature variations in the ocean. These are of great importance to fisheries.

Freshwater fish migrations are usually shorter, typically from lake to stream or vice versa, for spawning purposes.

Continental shelf

The continental shelf is the extended perimeter of each continent and associated coastal plain. Much of the shelf was exposed during glacial periods, but is now submerged under relatively shallow seas (known as shelf seas) and gulfs, and was similarly submerged during other interglacial periods.

The continental margin, between the continental shelf and the abyssal plain, comprises a steep continental slope followed by the flatter continental rise.

Bay of Bengal

The Bay of Bengal, the largest bay in the world, forms the northeastern part of the Indian Ocean. Roughly triangular in shape, it is bordered mostly by India and Sri Lanka to the west, Bangladesh to the north, and Burma (Myanmar) and the Andaman and Nicobar Islands to the east.

The Bay of Bengal occupies an area of 2,172,000 km².

Storm surge

A storm surge is an offshore rise of water associated with a low pressure weather system, typically tropical cyclones and strong extratropical cyclones. Storm surges are caused primarily by high winds pushing on the ocean's surface. The wind causes the water to pile up higher than the ordinary sea level.

Tidal bore

A tidal bore is a tidal phenomenon in which the leading edge of the incoming tide forms a wave (or waves) of water that travel up a river or narrow bay against the direction of the river or bay's current. As such, it is a true tidal wave and not to be confused with a tsunami, which is a large ocean wave traveling primarily on the open ocean.

1. In physics, a _____ - also known as a stationary wave - is a wave that remains in a constant position.

    This phenomenon can occur because the medium is moving in the opposite direction to the wave, or it can arise in a stationary medium as a result of interference between two waves traveling in opposite directions. In the second case, for waves of equal amplitude traveling in opposing directions, there is on average no net propagation of energy.

    a. Standing wave
    b. Chepo expedition
    c. Golden Hind
    d. Storm surge

2. _____s are gravity waves that oscillate within, rather than on the surface of, a fluid medium. They are one of many types of wave motion in stratified fluids (another example being Lee waves). A simple example is a wave propagating on the interface between two fluids of different densities, such as oil and water.

    a. Intertropical Convergence Zone
    b. Island wake
    c. Isopycnic
    d. Internal wave

3. The _____ Period is the most recent of the three periods of the Cenozoic Era in the geologic time scale of the ICS. It follows the Neogene Period, spanning 2.588 ± 0.005 million years ago to the present. The _____ includes two geologic epochs: the Pleistocene and the Holocene. Research history

    The term _____ was proposed by Giovanni Arduino in 1759 for alluvial deposits in the Po river valley in northern Italy.

    a. Quaternary
    b. Riffle-pool sequence
    c. Sahara pump theory
    d. Sediment transport

4. The _____ (symbol P_S) is the epoch from 2,588,000 to 11,700 years BP that spans the world's recent period of repeated glaciations. The name _____ is derived from the Greek πλε?στος (pleistos 'most') and καιν?ς (kainos 'new').

    The _____ Epoch follows the Pliocene Epoch and is followed by the Holocene Epoch.

    a. Quaternary science
    b. Radioanalytical chemistry
    c. Angola Current
    d. Pleistocene

5. . _____ is the transport of energy by ocean surface waves, and the capture of that energy to do useful work -- for example, electricity generation, water desalination, or the pumping of water (into reservoirs). Machinery able to exploit _____ is generally known as a wave energy converter (WEC).

Visit Cram101.com for full Practice Exams

a. Wave farm
b. Wave power
c. Angola Current
d. Antarctic Circumpolar Current

1. a
2. d
3. a
4. d
5. b

You can take the complete Chapter Practice Test

for Chapter 1. PART I: Chapter 1 - Chapter 2
on all key terms, persons, places, and concepts.

Online 99 Cents

http://www.epub47.32.14042.1.cram101.com/

Use www.Cram101.com for all your study needs

including Cram101's online interactive problem solving labs in

chemistry, statistics, mathematics, and more.

CHAPTER OUTLINE: KEY TERMS, PEOPLE, PLACES, CONCEPTS

	Irish Sea
	North Sea
	Continental shelf
	Estuary
	Island arc
	Isostasy
	Passive margin
	Plate tectonics
	Sediment
	Terrigenous sediment
	Mudflat
	Weathering
	Canary Current
	Current
	Coastal erosion
	Abyssal plain
	Continental rise
	Pelagic sediments
	Submarine canyon

Tsunami

Turbidite

Turbidity current

Pleistocene

Quaternary

Tidal power

Kelvin wave

Benthic boundary layer

Settling

Algae

Algal mat

Shear velocity

Rossby wave

Backscatter

Langmuir circulation

Intertidal zone

Littoral zone

Irish Sea	The Irish Sea separates the islands of Ireland and Great Britain. It is connected to the Celtic Sea in the south by St George's Channel, and to the Atlantic Ocean in the north by the North Channel. Anglesey is the largest island within the Irish Sea, followed by the Isle of Man.
North Sea	The North Sea is a marginal sea of the Atlantic Ocean located between Great Britain, Scandinavia, Germany, the Netherlands, and Belgium. An epeiric (or 'shelf') sea on the European continental shelf, it connects to the ocean through the English Channel in the south and the Norwegian Sea in the north. It is more than 970 kilometres (600 mi) long and 580 kilometres (360 mi) wide, with an area of around 750,000 square kilometres (290,000 sq mi).
Continental shelf	The continental shelf is the extended perimeter of each continent and associated coastal plain. Much of the shelf was exposed during glacial periods, but is now submerged under relatively shallow seas (known as shelf seas) and gulfs, and was similarly submerged during other interglacial periods.
	The continental margin, between the continental shelf and the abyssal plain, comprises a steep continental slope followed by the flatter continental rise.
Estuary	An estuary is a partly enclosed coastal body of water with one or more rivers or streams flowing into it, and with a free connection to the open sea.
	Estuaries form a transition zone between river environments and ocean environments and are subject to both marine influences, such as tides, waves, and the influx of saline water; and riverine influences, such as flows of fresh water and sediment. The inflow of both seawater and freshwater provide high levels of nutrients in both the water column and sediment, making estuaries among the most productive natural habitats in the world.
Island arc	An island arc is a type of archipelago composed of a chain of volcanoes which alignment is arc-shaped, and which are situated parallel and close to a boundary between two converging tectonic plates.
	Most of these island arcs are formed as one oceanic tectonic plate subducts another one and, in most cases, produces magma at depth below the over-riding plate. However, this is only true for those island arcs that are part of the group of mountain belts which are called volcanic arcs, a term which is used when all the elements of the arc-shaped mountain belt are composed of volcanoes.
Isostasy	Isostasy is a term used in geology to refer to the state of gravitational equilibrium between the earth's lithosphere and asthenosphere such that the tectonic plates 'float' at an elevation which depends on their thickness and density. This concept is invoked to explain how different topographic heights can exist at the Earth's surface.

Passive margin	A passive margin is the transition between oceanic and continental crust which is not an active plate margin. It is constructed by sedimentation above an ancient rift, now marked by transitional crust. Continental rifting creates new ocean basins.
Plate tectonics	Plate tectonics is a scientific theory that describes the large scale motions of Earth's lithosphere. The theory builds on the concepts of continental drift, developed during the first decades of the 20th century, and accepted by the majority of the geoscientific community when the concepts of seafloor spreading were developed in the late 1950s and early 1960s. The lithosphere is broken up into tectonic plates.
Sediment	Sediment, a naturally occurring material that is broken down by processes of weathering and erosion, and is subsequently transported by the action of fluids such as wind, water, or ice, and/or by the force of gravity acting on the particle itself. Sediments are most often transported by water (fluvial processes), wind (aeolian processes) and glaciers. Beach sands and river channel deposits are examples of fluvial transport and deposition, though sediment also often settles out of slow-moving or standing water in lakes and oceans.
Terrigenous sediment	In oceanography, terrigenous sediments are those derived from the erosion of rocks on land; that is, that are derived from terrestrial environments.(Pinet, 79) Consisting of sand, mud, and silt carried to sea by rivers, their composition is usually related to their source rocks; deposition of these sediments is largely limited to the continental shelf.(Pinet, 79-83) Sources of terrigenous sediments include volcanoes, weathering of rocks, wind-blown dust, grinding by glaciers, and sediment carried by icebergs.
Mudflat	Mudflats are coastal wetlands that form when mud is deposited by tides or rivers. They are found in sheltered areas such as bays, bayous, lagoons, and estuaries. Mudflats may be viewed geologically as exposed layers of bay mud, resulting from deposition of estuarine silts, clays and marine animal detritus. Most of the sediment within a mudflat is within the intertidal zone, and thus the flat is submerged and exposed approximately twice daily.
Weathering	Weathering is the breaking down of rocks, soils and minerals as well as artificial materials through contact with the Earth's atmosphere, biota and waters. Weathering occurs in situ, or 'with no movement', and thus should not be confused with erosion, which involves the movement of rocks and minerals by agents such as water, ice, wind, and gravity.

Canary Current	The Canary Current is a wind-driven surface current that is part of the North Atlantic Gyre. This eastern boundary current branches south from the North Atlantic Current and flows southwest about as far as Senegal where it turns west and later joins the Atlantic North Equatorial Current. he Canary Islands.
Current	A current, in a river or stream, is the flow of water influenced by gravity as the water moves downhill to reduce its potential energy. The current varies spatially as well as temporally within the stream, dependent upon the flow volume of water, stream gradient, and channel geometrics. In tidal zones, the current in rivers and streams may reverse on the flood tide before resuming on the ebb tide.
Coastal erosion	Coastal erosion is the wearing away of land and the removal of beach or dune sediments by wave action, tidal currents, wave currents, or drainage . Waves, generated by storms, wind, or fast moving motor craft, cause coastal erosion, which may take the form of long-term losses of sediment and rocks, or merely the temporary redistribution of coastal sediments; erosion in one location may result in accretion nearby. The study of erosion and sediment redistribution is called 'coastal morphodynamics'.
Abyssal plain	An abyssal plain is an underwater plain on the deep ocean floor, usually found at depths between 3000 and 6000 metres. Lying generally between the foot of a continental rise and a mid-ocean ridge, abyssal plains cover more than 50% of the Earth's surface. They are among the flattest, smoothest and least explored regions on Earth.
Continental rise	The continental rise is an underwater feature found between the continental slope and the abyssal plain. This feature can be found all around the world, and it represents the final stage in the boundary between continents and the deepest part of the ocean. The environment in the continental rise is quite unique, and many oceanographers study it extensively in the hopes of learning more about the ocean and geologic history.
Pelagic sediments	Pelagic sediment or pelagite is a fine-grained sediment that has accumulated by the settling of particles through the water column to the ocean floor beneath the open ocean far from land. These particles consist primarily of either the microscopic, calcareous or siliceous shells of phytoplankton or zooplankton; clay-size siliciclastic sediment; or some mixture of these. Trace amounts of meteoric dust and variable amounts of volcanic ash occur within pelagic sediments.
Submarine canyon	A submarine canyon is a steep-sided valley cut into the sea floor of the continental slope, sometimes extending well onto the continental shelf. Some submarine canyons are found as extensions to large rivers; however most of them have no such association. Canyons cutting the continental slopes have been found at depths greater than 2 km below sea level.

Tsunami	A tsunami is a series of water waves caused by the displacement of a large volume of a body of water, typically an ocean or a large lake. Earthquakes, volcanic eruptions and other underwater explosions (including detonations of underwater nuclear devices), landslides, glacier calvings, meteorite impacts and other disturbances above or below water all have the potential to generate a tsunami.
	Tsunami waves do not resemble normal sea waves, because their wavelength is far longer.
Turbidite	Turbidite geological formations have their origins in turbidity current deposits, which are deposits from a form of underwater avalanche that are responsible for distributing vast amounts of clastic sediment into the deep ocean.
	Turbidites were first properly described by Arnold H. Bouma (1962), who studied deepwater sediments and recognized particular fining up intervals within deep water, fine grained shales, which were anomalous because they started at pebble conglomerates and terminated in shales.
	This was anomalous because within the deep ocean it had historically been assumed that there was no mechanism by which tractional flow could carry and deposit coarse-grained sediments into the abyssal depths.
Turbidity current	A turbidity current is a current of rapidly moving, sediment-laden water moving down a slope through water, or another fluid. The current moves because it has a higher density and turbidity than the fluid through which it flows. The driving force of a turbidity current is obtained from the sediment, which renders the turbid water heavier than the clear water above.
Pleistocene	The Pleistocene (symbol P_S) is the epoch from 2,588,000 to 11,700 years BP that spans the world's recent period of repeated glaciations. The name pleistocene is derived from the Greek πλε?στος (pleistos 'most') and καιν?ς (kainos 'new').
	The Pleistocene Epoch follows the Pliocene Epoch and is followed by the Holocene Epoch.
Quaternary	The Quaternary Period is the most recent of the three periods of the Cenozoic Era in the geologic time scale of the ICS. It follows the Neogene Period, spanning 2.588 ± 0.005 million years ago to the present. The Quaternary includes two geologic epochs: the Pleistocene and the Holocene. Research history
	The term Quaternary was proposed by Giovanni Arduino in 1759 for alluvial deposits in the Po river valley in northern Italy.

Tidal power	Tidal power is a form of hydropower that converts the energy of tides into useful forms of power - mainly electricity. The first study of large scale tidal power plants was by the US Federal Power Commission in 1924 which would have been located if built in the northern border area of the US state of Maine and the south eastern border area of the Canadian province of New Brunswick, with various dams, powerhouses and ship locks enclosing the Bay of Fundy and Passamaquoddy Bay . Nothing came of the study and it is unknown whether Canada had been approached about the study by the US Federal Power Commission.
Kelvin wave	A Kelvin wave is a wave in the ocean or atmosphere that balances the Earth's Coriolis force against a topographic boundary such as a coastline, or a waveguide such as the equator. A feature of a Kelvin wave is that it is non-dispersive, i.e., the phase speed of the wave crests is equal to the group speed of the wave energy for all frequencies. This means that it retains its shape in the alongshore direction over time.
Benthic boundary layer	The benthic boundary layer is the layer of water directly above the sediment at the bottom of a river, lake or sea. It is generated by the friction of the water moving over the surface of the substrate. The thickness of this zone is determined by many factors including the Coriolis force.
Settling	Settling is the process by which particulates settle to the bottom of a liquid and form a sediment. Particles that experience a force, either due to gravity or due to centrifugal motion will tend to move in a uniform manner in the direction exerted by that force. For gravity settling, this means that the particles will tend to fall to the bottom of the vessel, forming a slurry at the vessel base.
Algae	Algae are a large and diverse group of simple, typically autotrophic organisms, ranging from unicellular to multicellular forms, such as the giant kelps that grow to 65 meters in length. They are photosynthetic like plants, and 'simple' because their tissues are not organized into the many distinct organs found in land plants. The largest and most complex marine forms are called seaweeds.
Algal mat	An algal mat is a layer of usually filamentous algae on marine or fresh water soft bottoms. It may be considered one of many types of microbial mats. Algae and cyanobacteria are ubiquitous, often forming within the water column and settling to the bottom.
Shear velocity	Shear velocity, is a form by which a shear stress may be re-written in units of velocity. It is useful as a method in fluid mechanics to compare true velocities, such as the velocity of a flow in a stream, to a velocity that relates shear between layers of flow. Shear velocity is used to describe shear-related motion in moving fluids.

Rossby wave	Atmospheric Rossby waves are giant meanders in high-altitude winds that are a major influence on weather. They are not to be confused with oceanic Rossby waves, which move along the thermocline: that is, the boundary between the warm upper layer of the ocean and the cold deeper part of the ocean. Rossby waves are a subset of inertial waves.
Backscatter	In physics, backscatter is the reflection of waves, particles, or signals back to the direction they came from. It is a diffuse reflection due to scattering, as opposed to specular reflection like a mirror. Backscattering has important applications in astronomy, photography and medical ultrasonography.
Langmuir circulation	Langmuir circulation consists of a series of shallow, slow, counter-rotating vortices at the ocean's surface. These circulations are developed when a particular type of wind blows steadily over the sea surface. Irving Langmuir discovered this phenomenon after observing windrows of seaweed in the Sargasso Sea in 1938. Langmuir circulations usually circulate water with a depth of no more than 66 feet, which does not allow upwelling to bring nutrient-rich waters from the pycnocline-typically with a depth of more than 3000 feet- to the ocean surface.
Intertidal zone	The intertidal zone, is the area that is above water at low tide and under water at high tide (in other words, the area between tide marks). This area can include many different types of habitats, with many types of animals like starfish, sea urchins, and some species of coral. The well known area also includes steep rocky cliffs, sandy beaches, or wetlands (e.g., vast mudflats).
Littoral zone	The littoral zone is that part of a sea, lake or river that is close to the shore. In coastal environments the littoral zone extends from the high water mark, which is rarely inundated, to shoreline areas that are permanently submerged. It always includes this intertidal zone and is often used to mean the same as the intertidal zone.

1. An _____ is an underwater plain on the deep ocean floor, usually found at depths between 3000 and 6000 metres. Lying generally between the foot of a continental rise and a mid-ocean ridge, _____s cover more than 50% of the Earth's surface. They are among the flattest, smoothest and least explored regions on Earth.

   a. Abyssal zone
   b. Acoustic release
   c. Abyssal plain
   d. Acoustical oceanography

2. A _____ is a current of rapidly moving, sediment-laden water moving down a slope through water, or another fluid. The current moves because it has a higher density and turbidity than the fluid through which it flows. The driving force of a _____ is obtained from the sediment, which renders the turbid water heavier than the clear water above.

   a. Chepo expedition
   b. Turbidity current
   c. Wave shoaling
   d. Weddell Gyre

3. The _____ is the extended perimeter of each continent and associated coastal plain. Much of the shelf was exposed during glacial periods, but is now submerged under relatively shallow seas (known as shelf seas) and gulfs, and was similarly submerged during other interglacial periods.

   The continental margin, between the _____ and the abyssal plain, comprises a steep continental slope followed by the flatter continental rise.

   a. Continental shelf
   b. Current meter
   c. Current sea level rise
   d. Cuspate foreland

4. An _____ is a partly enclosed coastal body of water with one or more rivers or streams flowing into it, and with a free connection to the open sea.

   Estuaries form a transition zone between river environments and ocean environments and are subject to both marine influences, such as tides, waves, and the influx of saline water; and riverine influences, such as flows of fresh water and sediment. The inflow of both seawater and freshwater provide high levels of nutrients in both the water column and sediment, making estuaries among the most productive natural habitats in the world.

   a. Antelope of Boston
   b. Current meter
   c. Current sea level rise
   d. Estuary

5. _____, a naturally occurring material that is broken down by processes of weathering and erosion, and is subsequently transported by the action of fluids such as wind, water, or ice, and/or by the force of gravity acting on the particle itself.

   _____s are most often transported by water (fluvial processes), wind (aeolian processes) and glaciers. Beach sands and river channel deposits are examples of fluvial transport and deposition, though _____ also often settles out of slow-moving or standing water in lakes and oceans.

   a. Chepo expedition
   b. Sediment
   c. Mesoplates
   d. Non-volcanic passive margins

ANSWER KEY
Chapter 2. PART II: Chapter 3 - Chapter 4

---

1. c
2. b
3. a
4. d
5. b

You can take the complete Chapter Practice Test

for Chapter 2. PART II: Chapter 3 - Chapter 4
on all key terms, persons, places, and concepts.

Online 99 Cents

http://www.epub47.32.14042.2.cram101.com/

Use www.Cram101.com for all your study needs

including Cram101's online interactive problem solving labs in

chemistry, statistics, mathematics, and more.

CHAPTER OUTLINE: KEY TERMS, PEOPLE, PLACES, CONCEPTS

	Baltic Sea
	Pleistocene
	Estuary
	Surf zone
	Mudflat
	Tidal power
	Rossby wave
	Infragravity wave
	Wave height
	Sediment
	Kelvin wave
	Current
	Lagoon
	Sediment transport
	Wave power
	Rip current
	Coastal erosion
	North Sea
	Salinity

Chapter 3. PART III: Chapter 5 - Chapter 8

_____ Algal mat

_____ Settling

_____ Canary Current

_____ Van der Waals force

_____ Black Sea

_____ Brackish water

_____ Internal wave

_____ Pycnocline

_____ Stratification

_____ Tidal range

_____ Sedimentation

_____ Mangrove

_____ Continental rise

_____ Continental shelf

_____ Submarine canyon

_____ Turbidite

_____ Turbidity current

_____ Wadden Sea

_____ Barrier island

CHAPTER OUTLINE: KEY TERMS, PEOPLE, PLACES, CONCEPTS

_____ | Persian Gulf

_____ | Algae

_____ | Stromatolite

_____ | Bay of Bengal

_____ | Isostasy

_____ | Distributary

_____ | Seawater

_____ | Weathering

_____ | Storm surge

_____ | Coccolithophore

_____ | Irish Sea

_____ | Blake Plateau

_____ | Ocean current

_____ | Upwelling

_____ | Wind stress

_____ | Callianassa

_____ | Bioturbation

_____ | Dredging

_____ | Phosphorite

Thermocline

Baltic Sea	The Baltic Sea is a brackish mediterranean sea located in Northern Europe, from 53°N to 66°N latitude and from 20°E to 26°E longitude. It is bounded by the Scandinavian Peninsula, the mainland of Europe, and the Danish islands. It drains into the Kattegat by way of the Øresund, the Great Belt and the Little Belt.
Pleistocene	The Pleistocene (symbol P_s) is the epoch from 2,588,000 to 11,700 years BP that spans the world's recent period of repeated glaciations. The name pleistocene is derived from the Greek πλε? στος (pleistos 'most') and καιν?ς (kainos 'new').
	The Pleistocene Epoch follows the Pliocene Epoch and is followed by the Holocene Epoch.
Estuary	An estuary is a partly enclosed coastal body of water with one or more rivers or streams flowing into it, and with a free connection to the open sea.
	Estuaries form a transition zone between river environments and ocean environments and are subject to both marine influences, such as tides, waves, and the influx of saline water; and riverine influences, such as flows of fresh water and sediment. The inflow of both seawater and freshwater provide high levels of nutrients in both the water column and sediment, making estuaries among the most productive natural habitats in the world.
Surf zone	As ocean surface waves come closer to shore they break, forming the foamy, bubbly surface we call surf. The region of breaking waves defines the surf zone. After breaking in the surf zone, the waves (now reduced in height) continue to move in, and they run up onto the sloping front of the beach, forming an uprush of water called swash.
Mudflat	Mudflats are coastal wetlands that form when mud is deposited by tides or rivers. They are found in sheltered areas such as bays, bayous, lagoons, and estuaries. Mudflats may be viewed geologically as exposed layers of bay mud, resulting from deposition of estuarine silts, clays and marine animal detritus. Most of the sediment within a mudflat is within the intertidal zone, and thus the flat is submerged and exposed approximately twice daily.

Chapter 3. PART III: Chapter 5 - Chapter 8

Tidal power	Tidal power is a form of hydropower that converts the energy of tides into useful forms of power - mainly electricity. The first study of large scale tidal power plants was by the US Federal Power Commission in 1924 which would have been located if built in the northern border area of the US state of Maine and the south eastern border area of the Canadian province of New Brunswick, with various dams, powerhouses and ship locks enclosing the Bay of Fundy and Passamaquoddy Bay . Nothing came of the study and it is unknown whether Canada had been approached about the study by the US Federal Power Commission.
Rossby wave	Atmospheric Rossby waves are giant meanders in high-altitude winds that are a major influence on weather. They are not to be confused with oceanic Rossby waves, which move along the thermocline: that is, the boundary between the warm upper layer of the ocean and the cold deeper part of the ocean. Rossby waves are a subset of inertial waves.
Infragravity wave	Infragravity waves are surface gravity waves with frequencies lower than the wind waves - consisting of both wind sea and swell - so corresponding with the part of the wave spectrum lower than the frequencies directly generated by forcing through the wind. Infragravity waves consist, among others, of long-period oceanic waves generated along continental coastlines by nonlinear wave interactions of storm-forced shoreward-propagating ocean swells. These differ from normal oceanic gravity waves, which are created by wind acting on the surface of the sea.
Wave height	In fluid dynamics, the wave height of a surface wave is the difference between the elevations of a crest and a neighbouring trough. Wave height is a term used by mariners, as well as in coastal, ocean and naval engineering. At sea, the term significant wave height is used as a means to introduce a well-defined and standardized statistic to denote the characteristic height of the random waves in a sea state.
Sediment	Sediment, a naturally occurring material that is broken down by processes of weathering and erosion, and is subsequently transported by the action of fluids such as wind, water, or ice, and/or by the force of gravity acting on the particle itself. Sediments are most often transported by water (fluvial processes), wind (aeolian processes) and glaciers.

Kelvin wave	A Kelvin wave is a wave in the ocean or atmosphere that balances the Earth's Coriolis force against a topographic boundary such as a coastline, or a waveguide such as the equator. A feature of a Kelvin wave is that it is non-dispersive, i.e., the phase speed of the wave crests is equal to the group speed of the wave energy for all frequencies. This means that it retains its shape in the alongshore direction over time.
Current	A current, in a river or stream, is the flow of water influenced by gravity as the water moves downhill to reduce its potential energy. The current varies spatially as well as temporally within the stream, dependent upon the flow volume of water, stream gradient, and channel geometrics. In tidal zones, the current in rivers and streams may reverse on the flood tide before resuming on the ebb tide.
Lagoon	A lagoon is a body of shallow sea water or brackish water separated from the sea by some form of barrier. The EU's habitat directive defines lagoons as 'expanses of shallow coastal salt water, of varying salinity or water volume, wholly or partially separated from the sea by sand banks or shingle, or, less frequently, by rocks. Salinity may vary from brackish water to hypersalinity depending on rainfall, evaporation and through the addition of fresh seawater from storms, temporary flooding by the sea in winter or tidal exchange'.
Sediment transport	Sediment transport is the movement of solid particles (sediment), typically due to a combination of the force of gravity acting on the sediment, and/or the movement of the fluid in which the sediment is entrained. An underanding of sediment transport is typically used in natural syems, where the particles are claic rocks (sand, gravel, boulders, etc)., mud, or clay; the fluid is air, water, or ice; and the force of gravity acts to move the particles due to the sloping surface on which they are reing. Sediment transport due to fluid motion occurs in rivers, the oceans, lakes, seas, and other bodies of water, due to currents and tides; in glaciers as they flow, and on terrerial surfaces under the influence of wind.
Wave power	Wave power is the transport of energy by ocean surface waves, and the capture of that energy to do useful work -- for example, electricity generation, water desalination, or the pumping of water (into reservoirs). Machinery able to exploit wave power is generally known as a wave energy converter (WEC).
Rip current	A rip current, commonly referred to by the misnomer rip tide or simply a rip, is a strong channel of water flowing seaward from near the shore, typically through the surf line. Typical flow is at 0.5 metres per second (1-2 feet per second), and can be as fast as 2.5 metres per second (8 feet per second). They can move to different locations on a beach break, up to tens of metres (a few hundred feet) a day.
Coastal erosion	Coastal erosion is the wearing away of land and the removal of beach or dune sediments by wave action, tidal currents, wave currents, or drainage .

Waves, generated by storms, wind, or fast moving motor craft, cause coastal erosion, which may take the form of long-term losses of sediment and rocks, or merely the temporary redistribution of coastal sediments; erosion in one location may result in accretion nearby. The study of erosion and sediment redistribution is called 'coastal morphodynamics'.

North Sea	The North Sea is a marginal sea of the Atlantic Ocean located between Great Britain, Scandinavia, Germany, the Netherlands, and Belgium. An epeiric (or 'shelf') sea on the European continental shelf, it connects to the ocean through the English Channel in the south and the Norwegian Sea in the north. It is more than 970 kilometres (600 mi) long and 580 kilometres (360 mi) wide, with an area of around 750,000 square kilometres (290,000 sq mi).
Salinity	Salinity is the saltiness or dissolved salt content of a body of water. It is a general term used to describe the levels of different salts such as sodium chloride, magnesium and calcium sulfates, and bicarbonates. Salinity in Australian English and North American English may also refer to the salt content of soil .
Algal mat	An algal mat is a layer of usually filamentous algae on marine or fresh water soft bottoms. It may be considered one of many types of microbial mats. Algae and cyanobacteria are ubiquitous, often forming within the water column and settling to the bottom.
Settling	Settling is the process by which particulates settle to the bottom of a liquid and form a sediment. Particles that experience a force, either due to gravity or due to centrifugal motion will tend to move in a uniform manner in the direction exerted by that force. For gravity settling, this means that the particles will tend to fall to the bottom of the vessel, forming a slurry at the vessel base.
Canary Current	The Canary Current is a wind-driven surface current that is part of the North Atlantic Gyre. This eastern boundary current branches south from the North Atlantic Current and flows southwest about as far as Senegal where it turns west and later joins the Atlantic North Equatorial Current. he Canary Islands.
Van der Waals force	The term includes:•force between two permanent dipoles (Keesom force)•force between a permanent dipole and a corresponding induced dipole (Debye force)•force between two instantaneously induced dipoles (London dispersion force) It is also sometimes used loosely as a synonym for the totality of intermolecular forces. Van der Waals forces are relatively weak compared to normal chemical bonds, but play a fundamental role in fields as diverse as supramolecular chemistry, structural biology, polymer science, nanotechnology, surface science, and condensed matter physics. Van der Waals forces define the chemical character of many organic compounds.

Black Sea	The Black Sea is bounded by Europe, Anatolia and the Caucasus and is ultimately connected to the Atlantic Ocean via the Mediterranean and the Aegean seas and various straits. The Bosphorus strait connects it to the Sea of Marmara, and the strait of the Dardanelles connects that sea to the Aegean Sea region of the Mediterranean. These waters separate eastern Europe and western Asia.
Brackish water	Water salinity based on dissolved salts in parts per thousand (ppt) Brackish water is water that has more salinity than fresh water, but not as much as seawater. It may result from mixing of seawater with fresh water, as in estuaries, or it may occur in brackish fossil aquifers. The word comes from the Middle Dutch root 'brak,' meaning 'salty'.
Internal wave	Internal waves are gravity waves that oscillate within, rather than on the surface of, a fluid medium. They are one of many types of wave motion in stratified fluids (another example being Lee waves). A simple example is a wave propagating on the interface between two fluids of different densities, such as oil and water.
Pycnocline	A pycnocline is the cline or layer where the density gradient ($\partial p/\partial z$) is greatest within a body of water. An ocean current is generated by the forces such as breaking waves, terms of temperature and salinity differences, wind, Coriolis effect, and tides caused by the gravitational pull of the Moon and the Sun. In addition, the physical properties in pycnocline driven by density gradients also affect the flows and vertical profiles in the ocean.
Stratification	Water stratification occurs when water of high and low salinity (halocline), oxygenation (chemocline), density (pycnocline), temperature (thermocline), forms layers that act as barriers to water mixing.
Tidal range	The tidal range is the vertical difference between the high tide and the succeeding low tide. Tides are the rise and fall of sea levels caused by the combined effects of the gravitational forces exerted by the Moon and the Sun and the rotation of the Earth. The tidal range is not constant, but changes depending on where the sun and the moon are.
Sedimentation	Sedimentation is the tendency for particles in suspension to settle out of the fluid in which they are entrained, and come to rest against a barrier. This is due to their motion through the fluid in response to the forces acting on them: these forces can be due to gravity, centrifugal acceleration or electromagnetism. In geology sedimentation is often used as the polar opposite of erosion, i.e., the terminal end of sediment transport.
Mangrove	Mangroves are various kinds of trees up to medium height and shrubs that grow in saline coastal sediment habitats in the tropics and subtropics - mainly between latitudes 25° N and 25° S.

The remaining mangrove forest areas of the world in 2000 was 53,190 square miles (137, 760 km²) spanning to 118 countries and territories The word is used in at least three senses: (1) most broadly to refer to the habitat and entire plant assemblage or mangal, for which the terms mangrove forest biome, mangrove swamp and mangrove forest are also used, (2) to refer to all trees and large shrubs in the mangrove swamp, and (3) narrowly to refer to the mangrove family of plants, the Rhizophoraceae, or even more specifically just to mangrove trees of the genus Rhizophora. The term 'mangrove' comes to English from Spanish (perhaps by way of Portuguese), and is of Caribbean origin, likely Taíno. It was earlier 'mangrow', but this was corrupted via folk etymology influence of 'grove'.

Continental rise	The continental rise is an underwater feature found between the continental slope and the abyssal plain. This feature can be found all around the world, and it represents the final stage in the boundary between continents and the deepest part of the ocean. The environment in the continental rise is quite unique, and many oceanographers study it extensively in the hopes of learning more about the ocean and geologic history.
Continental shelf	The continental shelf is the extended perimeter of each continent and associated coastal plain. Much of the shelf was exposed during glacial periods, but is now submerged under relatively shallow seas (known as shelf seas) and gulfs, and was similarly submerged during other interglacial periods. The continental margin, between the continental shelf and the abyssal plain, comprises a steep continental slope followed by the flatter continental rise.
Submarine canyon	A submarine canyon is a steep-sided valley cut into the sea floor of the continental slope, sometimes extending well onto the continental shelf. Some submarine canyons are found as extensions to large rivers; however most of them have no such association. Canyons cutting the continental slopes have been found at depths greater than 2 km below sea level.
Turbidite	Turbidite geological formations have their origins in turbidity current deposits, which are deposits from a form of underwater avalanche that are responsible for distributing vast amounts of clastic sediment into the deep ocean. Turbidites were first properly described by Arnold H. Bouma (1962), who studied deepwater sediments and recognized particular fining up intervals within deep water, fine grained shales, which were anomalous because they started at pebble conglomerates and terminated in shales.

Turbidity current	A turbidity current is a current of rapidly moving, sediment-laden water moving down a slope through water, or another fluid. The current moves because it has a higher density and turbidity than the fluid through which it flows. The driving force of a turbidity current is obtained from the sediment, which renders the turbid water heavier than the clear water above.
Wadden Sea	The Wadden Sea is an intertidal zone in the southeastern part of the North Sea. It lies between the coast of northwestern continental Europe and the range of Frisian Islands, forming a shallow body of water with tidal flats and wetlands. It is rich in biological diversity.
Barrier island	Barrier Islands, a coastal landform and a type of barrier system, are relatively narrow strips of sand that parallel the mainland coast. They usually occur in chains, consisting of anything from a few islands to more than a dozen. Excepting the tidal inlets that separate the islands, a barrier chain may extend uninterrupted for over a hundred kilometers, the longest and widest being Padre Island.
Persian Gulf	The Persian Gulf, in Western Asia, is an extension of the Indian Ocean located between Iran (Persia) and the Arabian Peninsula.

The Persian Gulf was the focus of the 1980-1988 Iran-Iraq War, in which each side attacked the other's oil tankers. In 1991, the Persian Gulf again was the background for what was called the 'Persian Gulf War' or the 'Gulf War' when Iraq invaded Kuwait and was subsequently pushed back, despite the fact that this conflict was primarily a land conflict. |
| Algae | Algae are a large and diverse group of simple, typically autotrophic organisms, ranging from unicellular to multicellular forms, such as the giant kelps that grow to 65 meters in length. They are photosynthetic like plants, and 'simple' because their tissues are not organized into the many distinct organs found in land plants. The largest and most complex marine forms are called seaweeds. |
| Stromatolite | Stromatolites are layered accretionary structures formed in shallow water by the trapping, binding and cementation of sedimentary grains by biofilms of microorganisms, especially cyanobacteria (commonly known as blue-green algae). They include some of the most ancient records of life on Earth. |
| Bay of Bengal | The Bay of Bengal, the largest bay in the world, forms the northeastern part of the Indian Ocean. Roughly triangular in shape, it is bordered mostly by India and Sri Lanka to the west, Bangladesh to the north, and Burma (Myanmar) and the Andaman and Nicobar Islands to the east.

The Bay of Bengal occupies an area of 2,172,000 km². |

Isostasy	Isostasy is a term used in geology to refer to the state of gravitational equilibrium between the earth's lithosphere and asthenosphere such that the tectonic plates 'float' at an elevation which depends on their thickness and density. This concept is invoked to explain how different topographic heights can exist at the Earth's surface. When a certain area of lithosphere reaches the state of isostasy, it is said to be in isostatic equilibrium.
Distributary	A distributary is a stream that branches off and flows away from a main stream channel. They are a common feature of river deltas. The phenomenon is known as river bifurcation.
Seawater	Seawater is water from a sea or ocean. On average, seawater in the world's oceans has a salinity of about 3.5% (35 g/L, or 599 mM). This means that every kilogram (roughly one litre by volume) of seawater has approximately 35 grams (1.2 oz) of dissolved salts (predominantly sodium (Na^+) and chloride (Cl^-) ions).
Weathering	Weathering is the breaking down of rocks, soils and minerals as well as artificial materials through contact with the Earth's atmosphere, biota and waters. Weathering occurs in situ, or 'with no movement', and thus should not be confused with erosion, which involves the movement of rocks and minerals by agents such as water, ice, wind, and gravity. Two important classifications of weathering processes exist - physical and chemical weathering.
Storm surge	A storm surge is an offshore rise of water associated with a low pressure weather system, typically tropical cyclones and strong extratropical cyclones. Storm surges are caused primarily by high winds pushing on the ocean's surface. The wind causes the water to pile up higher than the ordinary sea level.
Coccolithophore	Coccolithophores are single-celled algae, protists and phytoplankton belonging to the division of haptophytes. They are distinguished by special calcium carbonate plates (or scales) of uncertain function called coccoliths (calcareous nanoplankton), which are important microfossils. Coccolithophores are almost exclusively marine and are found in large numbers throughout the surface euphotic zone of the ocean.
Irish Sea	The Irish Sea separates the islands of Ireland and Great Britain. It is connected to the Celtic Sea in the south by St George's Channel, and to the Atlantic Ocean in the north by the North Channel. Anglesey is the largest island within the Irish Sea, followed by the Isle of Man.
Blake Plateau	The Blake Plateau lies between the continental shelf and the deep ocean basin 145 kilometers (90 miles) and 170 kilometers (106 miles) with a depth of about 500 meters (1,640 feet) inshore sloping to about 1,000 meters (3,281 feet) about 375 kilometers (233 miles) off shore where the Blake Escarpment drops steeply to the deep basin.

Blake Plateau lies off the southeastern United States coastal states of North Carolina, South Carolina, Georgia and Florida. The Blake Plateau, associated ridge and basin are named for the U.S. Coast and Geodetic Survey steamer George S. Blake, in service 1874-1905, that first used steel cable for oceanographic operations and pioneered deep ocean and Gulf Stream exploration.

Ocean current

An ocean current is a continuous, directed movement of ocean water generated by the forces acting upon this mean flow, such as breaking waves, wind, Coriolis effect, cabbeling, temperature and salinity differences and tides caused by the gravitational pull of the Moon and the Sun. Depth contours, shoreline configurations and interaction with other currents influence a current's direction and strength.

Ocean currents can flow for great distances, and together they create the great flow of the global conveyor belt which plays a dominant part in determining the climate of many of the Earth's regions.

Upwelling

Upwelling is an oceanographic phenomenon that involves wind-driven motion of dense, cooler, and usually nutrient-rich water towards the ocean surface, replacing the warmer, usually nutrient-depleted surface water. The increased availability in upwelling regions results in high levels of primary productivity and thus fishery production. Approximately 25% of the total global marine fish catches come from five upwellings that occupy only 5% of the total ocean area.

Wind stress

In physical oceanography and fluid dynamics, the wind stress is the shear stress exerted by the wind on the surface of large bodies of water - such as oceans, seas, estuaries and lakes. It is the force component parallel to the surface, per unit area, as applied by the wind on the water surface. The wind stress is affected by the wind speed, the shape of the wind waves and the atmospheric stratification.

Callianassa

Callianassa is a genus of mud shrimps, in the family Callianassidae. Three of the species in this genus (C. candida, C. tyrrhena and C. whitei) have been split off into a new genus, Pestarella, while others such as Callianassa filholi have been moved to Biffarius.

Bioturbation

In oceanography, limnology, pedology, geology (especially geomorphology and sedimentology), and archaeology, bioturbation is the displacement and mixing of sediment particles (i.e. sediment reworking) and solutes (i.e. bio-irrigation) by fauna (animals) or flora (plants). The mediators of bioturbation are typically annelid worms (e.g. polychaetes, oligochaetes), bivalves (e.g. mussels, clams), gastropods, holothurians, or any other infaunal or epifaunal organisms. Faunal activities, such as burrowing, ingestion and defecation of sediment grains, construction and maintenance of galleries, and infilling of abandoned dwellings, displace sediment grains and mix the sediment matrix.

Dredging	Dredging is an excavation activity or operation usually carried out at least partly underwater, in shallow seas or fresh water areas with the purpose of gathering up bottom sediments and disposing of them at a different location. This technique is often used to keep waterways navigable.
	It is also used as a way to replenish sand on some public beaches, where too much sand has been lost because of coastal erosion.
Phosphorite	Phosphorite is a non-detrital sedimentary rock which contains high amounts of phosphate bearing minerals. The phosphate content of phosphorite is at least 15 to 20% which is a large enrichment over the typical sedimentary rock content of less than 0.2%. The phosphate is present as fluorapatite $Ca_5(PO_4)_3F$ (CFA) typically in cryptocrystalline masses (grain sizes < 1 µm) referred to as collophane.
Thermocline	A thermocline is a thin but distinct layer in a large body of fluid (e.g. water, such as an ocean or lake, or air, such as an atmosphere), in which temperature changes more rapidly with depth than it does in the layers above or below. In the ocean, the thermocline may be thought of as an invisible blanket which separates the upper mixed layer from the calm deep water below. Depending largely on season, latitude and turbulent mixing by wind, thermoclines may be a semi-permanent feature of the body of water in which they occur, or they may form temporarily in response to phenomena such as the radiative heating/cooling of surface water during the day/night.

CHAPTER QUIZ: KEY TERMS, PEOPLE, PLACES, CONCEPTS

1. A _____ is a body of shallow sea water or brackish water separated from the sea by some form of barrier. The EU's habitat directive defines _____s as 'expanses of shallow coastal salt water, of varying salinity or water volume, wholly or partially separated from the sea by sand banks or shingle, or, less frequently, by rocks. Salinity may vary from brackish water to hypersalinity depending on rainfall, evaporation and through the addition of fresh seawater from storms, temporary flooding by the sea in winter or tidal exchange'.

   a. Lagoon
   b. Natural arch
   c. Sea cave
   d. Stack

2. . Water _____ occurs when water of high and low salinity (halocline), oxygenation (chemocline), density (pycnocline), temperature (thermocline), forms layers that act as barriers to water mixing.

   a. Stratification
   b. Wave turbulence
   c. World Ocean Database Project

Chapter 3. PART III: Chapter 5 - Chapter 8

Visit Cram101.com for full Practice Exams

3. _____ is the saltiness or dissolved salt content of a body of water. It is a general term used to describe the levels of different salts such as sodium chloride, magnesium and calcium sulfates, and bicarbonates. _____ in Australian English and North American English may also refer to the salt content of soil .

   a. Sea foam
   b. Salinity
   c. Seawater
   d. Sediment trap

4. The _____ is the vertical difference between the high tide and the succeeding low tide. Tides are the rise and fall of sea levels caused by the combined effects of the gravitational forces exerted by the Moon and the Sun and the rotation of the Earth. The _____ is not constant, but changes depending on where the sun and the moon are.

   a. Pycnocline
   b. Redfield ratio
   c. Tidal range
   d. Terrigenous sediment

5. As ocean surface waves come closer to shore they break, forming the foamy, bubbly surface we call surf. The region of breaking waves defines the _____. After breaking in the _____, the waves (now reduced in height) continue to move in, and they run up onto the sloping front of the beach, forming an uprush of water called swash.

   a. Sverdrup balance
   b. Tasman Outflow
   c. Thermocline
   d. Surf zone

ANSWER KEY
Chapter 3. PART III: Chapter 5 - Chapter 8

---

1. a

2. a

3. b

4. c

5. d

You can take the complete Chapter Practice Test

for Chapter 3. PART III: Chapter 5 - Chapter 8
on all key terms, persons, places, and concepts.

Online 99 Cents

http://www.epub47.32.14042.3.cram101.com/

Use www.Cram101.com for all your study needs

including Cram101's online interactive problem solving labs in

chemistry, statistics, mathematics, and more.

Other Cram101 e-Books and Tests

Want More?
Cram101.com...

Cram101.com provides the outlines and highlights of your
textbooks, just like this e-StudyGuide, but also gives you the
PRACTICE TESTS, and other exclusive study tools for all of your
textbooks.

Learn More. *Just click*
http://www.cram101.com/

CPSIA information can be obtained at www.ICGtesting.com
Printed in the USA
LVOW091208240912

300060LV00001B/31/P